*Dedicated to
JC, Megan and Zach*

CECELIA TAKES FLIGHT

from the drawing board to the sky

By Noreen Anne

Illustrated by Nithini Wathsala

ISBN: 979-8-9906498-5-9

Library of Congress Control Number: 2025906445

Copyright © 2025, Little Angels Book Club Ltd.

All rights reserved. No part of this publication, text or illustrations may be reproduced, distributed, or transmitted in any form or means without prior permission of the publisher except in the event of brief quotations and in a limited manner in articles or reviews.

Contact: Noreen Anne, noreenanne4@gmail.com

CECELIA TAKES FLIGHT

from the
drawing board
to the sky

By Noreen Anne

CECELIA'S DREAM
was as big as the sky.

She loved painting pictures of airplanes, but more than anything, Cecelia wanted a *Hercules model airplane*—just like the one her Mommy flies on super important missions.

Sammi Sylvester made money selling lemonade and cupcakes. Even Mr. Grizzle who never buys anything, came back for seconds.

Cecelia painted a sign and made up a jingle:

Artwork for sale from me to you

Helps my special dream come true

They walked to Mrs. Summer's, Mr. Mathison's and Mrs. Addison's house.

Up the block and down the block, but Cecelia couldn't rest until everything was just right.

Later that evening . . .

Drifting off to sleep,
Cecelia could see herself flying—
soaring above the shimmering night sky.
She whispered to the wind.

Knock! Knock! Knock!
"Time to get up!"
Daddy shouted.

Cecelia bolted out of bed, and ran downstairs.

She set up the cash register as Daddy stepped back to admire the paintings.

But what if people laugh at my drawings or think my pictures are silly?

Just then, Mrs. Summers said, "I'll take two."

Mr. Mathison bought three,

and Mrs. Addison bought five pieces.

*What if nobody **ELSE** buys anything?* Cecelia worried.

But she stood tall, held her chin up and smiled.

Cecelia counted her earnings but she STILL needed $15.00.

Cecelia didn't move.
Her words wouldn't come out.
She took a deep breath
and didn't look away.

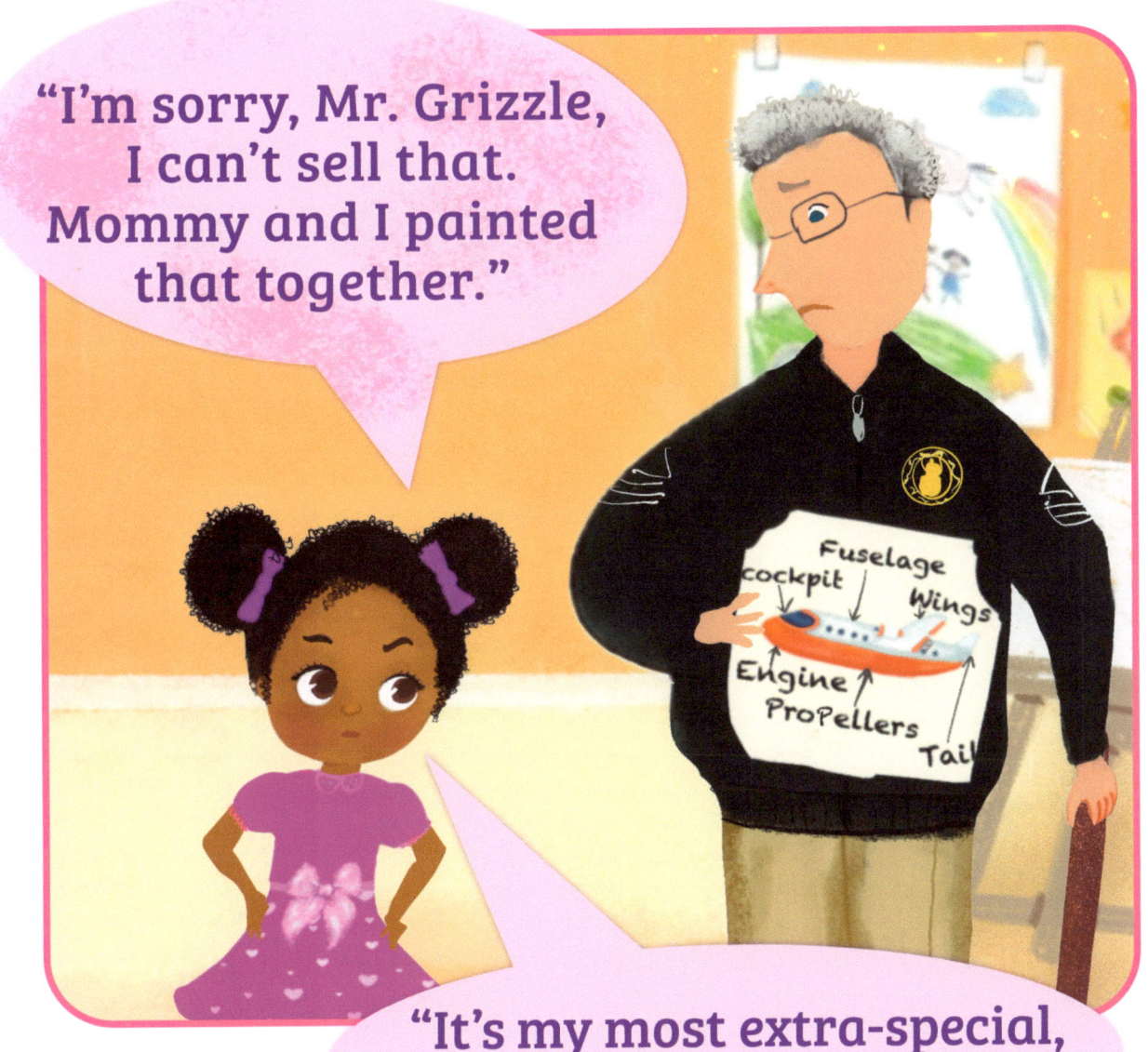

"I'm sorry, Mr. Grizzle, I can't sell that. Mommy and I painted that together."

"It's my most extra-special, keep-forever FAVORITE!"

"I'm saving for a model airplane just like the one my Mommy flies when she's away on special missions.

I need $15.00 more so I can buy it."

Just $15 more...

"I'm going to be a pilot too,
soaring above treetops
and racing with the wind!

But today,
thanks to you,
Mr. Grizzle,
it feels like
I have wings already—
like the stars
are whispering..."

"Cecelia, it's your turn now!"

The next day, skipping and dancing on their way to the store,

Cecelia sang—

Artwork for sale from me to you
Made my special dream come true

Help Cecelia
Find the Parts

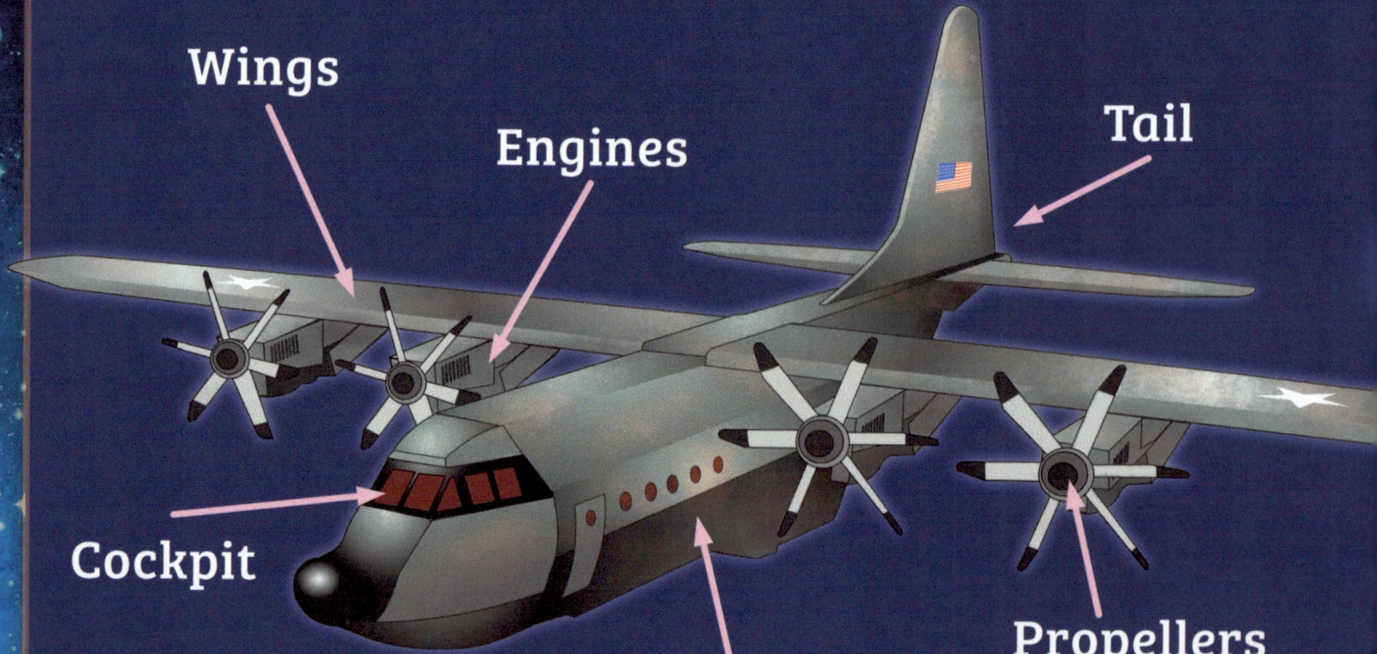

Wings

Engines

Tail

Cockpit

Propellers

Fuselage

The front area where the pilot sits and controls the plane.

The large flat parts on each side of the plane that helps it fly.

The powerful part of the plane generally mounted on the wings to help it fly.

The rear part of the plane that helps balance and stabilize the plane.

The main body of the airplane that holds the passengers and cargo.

The spinning blades at the front of the engines that push the airplane forward.

Meet the Author

Noreen Anne holds a Masters in Finance from DePaul University. Nowadays she prefers working with colored pencils and paint to weave together magical worlds. When she's not illustrating and crafting stories, she enjoys hiking outdoors and sharing her love of books during classroom visits.

If you enjoyed this book, please consider leaving a review online, sharing it with a friend or asking your local library to carry it. A portion of the net proceeds will be donated to the Zachary Tyler children's charity.

Illustrator

Nithini Wathsala lives in Sri Lanka and enjoys creating magical and enchanting environments. When she's not illustrating, she enjoys traveling and watching movies.

Acknowledgments

To all the mothers and fathers serving away from home on important missions, thank you for your service.

To my husband, family, and cherished readers — your love and support mean the world to me.

To Talara Ruth, Deb Robertson, Illoguild, my gifted illustrator, Nithini Wathsala and brilliant book designer, Chris Hammond thank you for making this book a reality. Above all, to Jesus whose boundless love lifts me up every day. "Thank you," feels far too inadequate when I owe him my life.

www.ingramcontent.com/pod-product-compliance
Lightning Source LLC
Chambersburg PA
CBHW041136130526
44582CB00031B/141